This book belongs to:

©2021 ABC Publishing

# The Solar System

# The Solar System

The solar system includes the Sun and everything that orbits (circles) around it: 8 planets, satellites, millions of asteroids, comets, dust and interplanetary gases.

In order of distance from the Sun, the eight planets are: Mercury, Venus, Earth, Mars, Jupiter, Saturn, Uranus and Neptune.

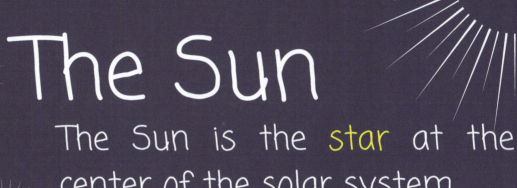

# The Sun

The Sun is the star at the center of the solar system.

It is a hot ball of gases that produces great amounts of energy.

Life on Earth would not be possible without the light and heat produced by the Sun.

All of the objects in the solar systems, such as planets and satellites orbit, or travel, around the Sun.

The Sun is the largest object in the solar system.

# The Moon

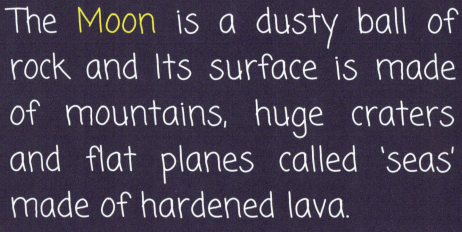

The Moon is a dusty ball of rock and Its surface is made of mountains, huge craters and flat planes called 'seas' made of hardened lava.

The Moon is Earth's only natural satellite. A satellite is an object that orbits a planet. Although the Moon shines bright in the night sky, it doesn't produce its own light, but it reflects light from the Sun.

# Asteroids

Asteroids are small, rocky objects that orbit the Sun.

Asteroids have jagged and irregular shapes. They range in size from nearly 600 miles (950 kilometers) across to chunky rocks less than half a mile (1 kilometer) wide.

The majority of asteroids orbit the Sun in a ring called the asteroid belt. The asteroid belt is located between the planets Mars and Jupiter.

# Comets

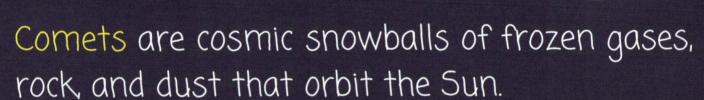

Comets are cosmic snowballs of frozen gases, rock, and dust that orbit the Sun.

When frozen, they are the size of a small town.

When a comet's orbit brings it close to the Sun, it heats up and spews dust and gases into a giant glowing head larger than most planets. The dust and gases then form a tail that stretches away from the Sun for millions of miles.

# Days and Years

We don't feel it, but the Earth is always spinning. It spins around an imaginary line running from the North to the South Pole, which we call its axis.

A 'day' is the amount of time it takes for a planet to make one complete spin.

As well as spinning, the Earth is also travelling around the Sun.

A 'year' is the amount of time it takes for Earth to travel all of the way around the Sun and back to where it started.

All of the planets in our Solar System spin on their axis and travel around the Sun. This means that they have days and years.

# Seasons

As we've said, the Earth spins around an *axis*. The Earth's axis is slightly *tilted*, though, at an angle of 23.5°. This means that, as the Earth orbits the Sun, at certain times of the year parts of the Earth are tilted towards the Sun, making the weather warmer.

So when a part of the Earth is tilted towards the Sun, we have summer months, and when it's tilted away from the Sun we have winter.

# Planets

Planets are the largest objects orbiting the Sun. The planets of the solar system can be divided into 2 large groups:

- Inner planets are closer to the Sun, small, warm, and composed largely of rock and metals. They include: Earth, Mercury, Venus and Mars.
- Outer planets are further away from the Sun. They are large and cold, and mostly composed of gases. They include: Jupiter, Saturn, Uranus and Neptune.

# Mercury

Mercury is the smallest planet in our solar system. It's just a little bigger than Earth's moon..

It is the closest planet to the sun, but it's actually not the hottest.

It has a solid surface that is covered with craters.

Mercury doesn't have an atmosphere,

A day on Mercury lasts 59 Earth days.

A year on Mercury lasts 88 Earth days.

# Venus

Venus is the second planet from the Sun and it has a thick atmosphere. that traps heat and makes it the hottest planet in the Solar System.
It has mountains and volcanoes.
Venus spins in the opposite direction of Earth and most other planets.
A day on Venus lasts 243 Earth days.
A year on Venus lasts 225 Earth days.

# Earth

Our planet Earth has a solid surface with mountains, valleys, canyons and so much more.

Water covers 70% of Earth's surface.

Earth's atmosphere keeps the planet warm so living things like us can be there, and it has plenty of oxygen for us to breathe..

Earth has one Moon.

A day on Earth lasts a little under 24 hours.

One year on Earth lasts 365.25 days. We add that extra 0.25 to our calendar every four years (leap year).

Earth is the only planet that has not been named after a Roman god or goddess.

Earth is also the only planet known to support life.

# Mars

Mars is a cold desert world. It is sometimes called the Red Planet. It's red because of rusty iron in the ground.
Mars has seasons, polar ice caps, volcanoes, canyons, and weather. It has a very thin atmosphere.
There are signs of ancient floods on Mars, but now water mostly exists in icy dirt and thin clouds.
Mars has two moons. Their names are Phobos and Deimos.
One day on Mars lasts 24.6 hours. One year on Mars is 687 Earth days.

# Jupiter

Jupiter is the biggest planet in our solar system. It is covered in swirling cloud stripes.

It has big storms like the Great Red Spot, which has been going for hundreds of years.

Jupiter doesn't have a solid surface, but it may have a solid inner core about the size of Earth.

Jupiter also has rings, but they're too faint to see very well.

Jupiter has 79 confirmed moons.

One day on Jupiter lasts 10 hours.

One year on Jupiter is the same as 11.8 Earth years.

# Saturn

Saturn isn't the only planet to have rings, but it definitely has the most beautiful ones. The 7 rings we see are made of groups of tiny ringlets that surround Saturn. They're made of chunks of ice and rock.

Like Jupiter, Saturn is mostly a ball of hydrogen and helium.

Saturn has 53 moons! It also has 29 unconfirmed moons we need to learn more about. One day on Saturn goes by in just 10.7 hours. One year on Saturn is the same as 29 Earth years.

# Uranus

Uranus is mostly made of flowing icy materials above a solid core.

Its atmosphere is made of hydrogen, helium and methane. The methane makes Uranus blue.

Uranus also has faint rings. The inner rings are narrow and dark. The outer rings are brightly colored and easier to see.

Like Venus, Uranus rotates in the opposite direction as most other planets. And unlike any other planet, Uranus rotates on its side.

Uranus has 27 known moons.

One day on Uranus lasts a little over 17 hours. One year on Uranus is the same as 84 years on Earth.

# Neptune

Neptune is dark, cold, and windy. Similar to Uranus, it is made of a thick fog of water, ammonia, and methane over a solid center.

Its atmosphere is made of hydrogen, helium, and methane. The methane gives Neptune the same blue color as Uranus.

Neptune has six rings, but they're very hard to see.

Neptune has 13 moons.

One day on Neptune goes by in 16 hours.

One year on Uranus is the same as 165 Earth years.

# Find the words

ASTEROID
COMET
EARTH
JUPITER
MARS
MERCURY
MOON
NEPTUNE
PLANET
SATELLITE
SATURN
STAR
SUN
URANUS
VENUS

```
S S R I S R E T I P U J Y D N
H U A U M M S K N O A U R I E
B G N T V Q W P W C V N U O P
H E Z A U T D V Y S W O C R T
V C Q B R R S K P P M W R E U
C P S S E U N M O O N L E T N
L G N Z D Z U Y C D X N M S E
J L Z L S A J J Y O A D C A C
C P F B C J P N O X L T S R O
X Y X Q K L U S V S V E A L M
R Z W M A S G Y T X Q D R S E
F Y A N A H F L J A N Y V Z T
O R E O Q M E M S U R Z B L N
S T I A B E A R T H G O Q S S
N H E T I L L E T A S Q B C E
```

# Name the planet

Check the answers on the first page of this book!

# Quiz Time

The Sun is:
A: a satellite
B: a star
C: a planet
D: a comet

Saturn is the only planet to have rings:
True ☐
False ☐

There are ………… planets in the Solar System.
A: 9
B: 6
C: 8
D: 7

# Quiz Time

Where in the Solar System are asteroids found?
_____
_____

The planets with rings are:
_____
_____

The planets without rings are:
_____
_____

Which planet has the most moons?
_____

# References

Bbc.co.uk. 2021. Time: Additional resources. [online] Available at: <https://www.bbc.co.uk/teach/terrific-scientific/KS2/z7mtbdm> [Accessed 13 August 2021].

Britannica Kids. n.d. Comet. [online] Available at: <https://kids.britannica.com/kids/article/comet/352987> [Accessed 12 August 2021].

Britannica Kids. n.d. Sun. [online] Available at: <https://kids.britannica.com/kids/article/Sun/353824> [Accessed 12 August 2021].

Focus Junior. 2021. Sistema Solare: i pianeti terrestri. [online] Available at: <https://www.focusjunior.it/scienza/spazio/pianeti/i-pianeti-del-sistema-solare/> [Accessed 12 August 2021].

NASA Science Solar System Exploration. n.d. Comets. [online] Available at: <https://solarsystem.nasa.gov/asteroids-comets-and-meteors/comets/overview/?page=0&per_page=40&order=name+asc&search=&condition_1=102%3Aparent_id&condition_2=comet%3Abody_type%3Ailike> [Accessed 12 August 2021].

NASA Science Space Place. 2019. All About the Planets. [online] Available at: <https://spaceplace.nasa.gov/planets/en/> [Accessed 12 August 2021].

NASA Science Space Place. 2021. What Is an Asteroid?. [online] Available at: <https://spaceplace.nasa.gov/asteroid/en/> [Accessed 12 August 2021].

National Geographic Kids. n.d. Facts about the Moon!. [online] Available at: <https://www.natgeokids.com/uk/discover/science/space/facts-about-the-moon/> [Accessed 12 August 2021].

Printed in the USA
CPSIA information can be obtained
at www.ICGtesting.com
LVHW060741171124
796845LV00025B/137